SOLUTIONS MANUAL FOR

INTRODUCTION TO

General, Organic, AND Biochemistry

SEVENTH EDITION

MORRIS HEIN
Mount San Antonio College

SCOTT PATTISON
Ball State University

SUSAN ARENA
University of Illinois at Urbana-Champaign

BROOKS/COLE

THOMSON LEARNING

Australia • Canada • Mexico • Singapore • Spain • United Kingdom • United States

BROOKS/COLE

THOMSON LEARNING

Sponsoring Editor: Marcus Boggs
Marketing Manager: Tom Ziolkowski
Editorial Assistant: Emily Levitan
Production Coordinator: Stephanie Andersen
Permissions Editor: Sue Ewing

Cover Design: Vernon T. Boes
Cover Photo: Ken Eward/BioGrafx
Print Buyer: Christopher Burnham
Printing and Binding: Webcom Ltd.

For more information about this or any other Brooks/Cole products, contact:
BROOKS/COLE
511 Forest Lodge Road
Pacific Grove, CA 93950 USA
www.brookscole.com
1-800-423-0563 (Thomson Learning Academic Resource Center)

Printed in Canada

10   9   8   7   6   5   4   3   2   1

ISBN: 0-534-38064-6